SRA

Connecting Math Concepts

Level C Student Assessment Book

COMPREHENSIVE EDITION

A DIRECT INSTRUCTION PROGRAM

McGraw Hill Education

Bothell, WA • Chicago, IL • Columbus, OH • New York, NY

mheducation.com/prek-12

Send all inquiries to:
McGraw-Hill Education
4400 Easton Commons
Columbus, OH 43219

ISBN: 978-0-02-103597-7
MHID: 0-02-103597-0

Printed in the United States of America.

17 18 19 20 LON 27 26 25 24 23

Mastery Test 1

Name _____

Part 1

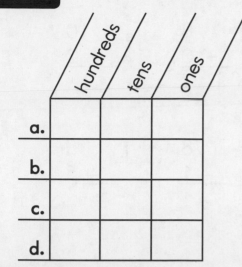

Part 2

a. $6 + 1 =$ _____

$6 + 2 =$ _____

b. $3 + 1 =$ _____

$3 + 2 =$ _____

c. $9 - 1 =$ _____

$9 - 2 =$ _____

d. $7 - 1 =$ _____

$7 - 2 =$ _____

Part 3 Write the missing numbers.

a. 10   9   __   __   __   __   4   __   __   __

b. 20   __   18   __   __   __   __   13   __   __   __

Part 4 Write 4 facts for each family.

a. 4 ——— 2 → 6

_____

_____

_____

_____

b. 6 ——— 3 → 9

_____

_____

_____

_____

Mastery Test 1

Name _____

Part 5

a.   4
    + 1

b.   7
    − 7

c.   6
    − 1

d.   8
    + 0

e. 10
    + 2

f.   6
    − 0

g.   9
    + 1

h.   3
    + 2

i.   8
    − 1

j.   5
    − 5

Part 6

a. 342

hundreds digit _____

ones digit _____

b. 50

tens digit _____

ones digit _____

c. 417

hundreds digit _____

tens digit _____

Part 7

a.   32
    + 12

b.   47
    + 30

c.   54
    + 21

Remedies

Name _____

Part A

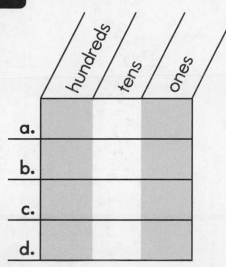

Part B

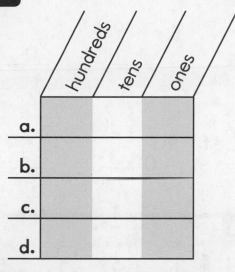

Part C

a. $3 + 1$	g. $9 + 1$
b. $3 + 2$	h. $9 + 2$
c. $6 + 1$	i. $7 + 1$
d. $6 + 2$	j. $7 + 2$
e. $4 + 1$	
f. $4 + 2$	

Part D

a. $6 - 1 = $____	g. $8 - 1 = $____
b. $6 - 2 = $____	h. $8 - 2 = $____
c. $4 - 1 = $____	i. $9 - 1 = $____
d. $4 - 2 = $____	j. $9 - 2 = $____
e. $10 - 1 = $____	k. $12 - 1 = $____
f. $10 - 2 = $____	l. $12 - 2 = $____

Part E

a. $\xrightarrow{\quad 4 \qquad 2 \quad} 6$

_____

_____

_____

Part F

a. $\xrightarrow{\quad 6 \qquad 3 \quad} 9$

_____

_____

_____

Remedies

Remedies CONTINUED

Name _____

Part G

a. 4
+ 3

b. 9
+ 1

c. 5
+ 2

d. 2
+ 3

e. 8
+ 3

f. 7
+ 0

g. 6
+ 2

h. 4
+ 3

i. 7
+ 3

j. 8
+ 1

k. 9
+ 2

l. 5
+ 3

Part H

a. 4
+ 2

b. 6
+ 0

c. 6
+ 2

d. 8
− 8

e. 7
− 0

f. 9
− 1

g. 5
+ 1

h. 9
− 0

i. 10
+ 0

j. 10
+ 2

k. 8
− 1

l. 7
− 7

Part I

a. 407

hundreds digit _____

tens digit _____

b. 37

tens digit _____

ones digit _____

c. 236

hundreds digit _____

ones digit _____

d. 910

tens digit _____

ones digit _____

Part J

a. 11
+ 48

b. 25
+ 12

c. 78
+ 20

d. 16
+ 31

Part K

a. 55
+ 32

b. 66
+ 32

c. 11
+ 33

d. 56
+ 22

Mastery Test 2

Name _____

Part 1

a. 9 – 1 = _____

b. 1 + 6 = _____

c. 4 – 1 = _____

d. 10 + 1 = _____

e. 9 – 8 = _____

f. 8 + 1 = _____

g. 8 – 7 = _____

h. 6 + 1 = _____

i. 10 – 1 = _____

j. 1 + 5 = _____

k. 6 – 5 = _____

l. 1 + 3 = _____

m. 8 – 1 = _____

n. 9 + 1 = _____

o. 3 – 2 = _____

Part 2 Write the fact. Then write the missing number.

a. 9 ———1——→ ___

_____

b. 6 ——————→ 8

_____

c. ——————2——→ 10

_____

d. 6 ———3——→ ___

_____

e. 10 ———2——→ ___

_____

f. ——————10——→ 11

_____

Part 3 Write the missing numbers.

a. 63 ___ ___ 60 ___ ___ ___ 56 ___

b. 100 90 ___ ___ ___ ___ 50 ___ ___ ___ ___ ___

Part 4 Write the number of cents for each row.

a.    _____ cents

b.   _____ cents

Mastery Test 2

Name _____

Part 5 Write the place-value addition for each number.

a. _____ = 16   b. _____ = 72

c. _____ = 61   d. _____ = 80

Part 6 Write the missing numbers.

a. 2 ___ ___ 8 ___ ___ ___ ___ ___ 20

b. 5 10 ___ ___ 25 ___ ___ ___ ___ ___

c. 9
___
27
___
___
54
___
___
90

Part 7 Work each problem.

a. 62   b. 52
  + 10   − 21

c. 78   d. 46
  − 12   + 30

Part 8 Write the answers.

a. 56 + 10 = _____   b. 80 + 10 = _____   c. 17 + 10 = _____

Connecting Math Concepts

Mastery Test 2

Name _____

Part 9 Write the fact. Then write the missing number in the family.

a. $6 \xrightarrow{\ T\ } 7$

_____

b. $M \xrightarrow{\ 2\ } 10$

_____

c. $8 \xrightarrow{\ 3\ } E$

_____

Part 10 Write the number of cents for each row.

a. _____ cents

b. _____ cents

Part 11 Write the answer.

a. $6 + 9 =$ _____   b. $8 + 9 =$ _____   c. $5 + 9 =$ _____

d. $3 + 9 =$ _____   e. $7 + 9 =$ _____   f. $4 + 9 =$ _____

Part 12 Work each problem.

a.
$$\begin{array}{r} 6 \\ 1 \\ +\ 3 \\ \hline \end{array}$$

b.
$$\begin{array}{r} 7 \\ 3 \\ +\ 4 \\ \hline \end{array}$$

c.
$$\begin{array}{r} 5 \\ 9 \\ +\ 1 \\ \hline \end{array}$$

d.
$$\begin{array}{r} 2 \\ 2 \\ +\ 2 \\ \hline \end{array}$$

Remedies

Name _____

Part A

a. $4 - 1 =$ _____

b. $7 - 6 =$ _____

c. $1 + 5 =$ _____

d. $10 - 9 =$ _____

e. $11 - 1 =$ _____

f. $9 - 8 =$ _____

g. $1 + 8 =$ _____

h. $3 - 2 =$ _____

i. $1 + 10 =$ _____

j. $9 - 1 =$ _____

Part B

a. $3 - 1 =$ _____

b. $1 + 2 =$ _____

c. $3 - 2 =$ _____

d. $6 - 1 =$ _____

e. $1 + 5 =$ _____

f. $9 - 8 =$ _____

g. $10 - 1 =$ _____

h. $1 + 10 =$ _____

i. $11 - 1 =$ _____

j. $1 + 3 =$ _____

k. $10 - 9 =$ _____

l. $4 - 1 =$ _____

m. $1 + 9 =$ _____

n. $6 - 5 =$ _____

o. $2 - 1 =$ _____

p. $1 + 4 =$ _____

q. $8 - 7 =$ _____

r. $7 - 1 =$ _____

s. $1 + 8 =$ _____

t. $4 - 3 =$ _____

u. $1 + 7 =$ _____

v. $8 - 1 =$ _____

w. $11 - 10 =$ _____

x. $5 - 4 =$ _____

y. $9 + 1 =$ _____

z. $5 - 1 =$ _____

A. $8 + 1 =$ _____

B. $7 - 6 =$ _____

C. $9 - 1 =$ _____

D. $1 + 6 =$ _____

Part C

a. $5 \longrightarrow 6$

b. $8 \quad 2 \longrightarrow$ _____

c. _____ $\xrightarrow{1} 10$

d. $6 \quad 2 \longrightarrow$ _____

Connecting Math Concepts

Remedies CONTINUED

Name _____

Part D

a. $6 \longrightarrow 8$

b. $\underline{\quad} \overset{1}{\longrightarrow} 5$

c. $7 \overset{2}{\longrightarrow} \underline{\quad}$

d. $\underline{\quad} \overset{2}{\longrightarrow} 10$

e. $5 \longrightarrow 7$

f. $9 \longrightarrow 11$

Part E Write the missing numbers.

72   71   ___   ___   68   ___   ___   ___

Part F Write the missing numbers.

a. 100   ___   ___   70   ___   ___   ___   ___   10

b. 83   82   ___   ___   78   ___   ___   ___

Part G

a.    cents

b.     cents

c.   cents

d.   cents

Remedies CONTINUED

Name _____

Part H Write the cents for each row.

a.    cents

b.   cents

c.   cents

Part I

a. _____ 37   b. _____ 60

c. _____ 12   d. _____ 10

Part J

a. _____ 16   b. _____ 20

c. _____ 10   d. _____ 96

Part K

a. 24   b. 52   c. 16   d. 46   e. 91
   + 33   – 10   + 13   + 32   – 20

Part L

a. 15   b. 36   c. 49   d. 57   e. 99
   + 23   – 20   – 12   + 31   – 22

Connecting Math Concepts

Remedies CONTINUED

Name _____

Part M

a. $40 + 10 =$ _____

b. $35 + 10 =$ _____

c. $17 + 10 =$ _____

d. $51 + 10 =$ _____

e. $89 + 10 =$ _____

Part N

a. $\underrightarrow{V \qquad 2} 9$

_____

b. $\underrightarrow{10 \qquad 5} D$

_____

c. $\underrightarrow{30 \qquad 7} F$

_____

d. $\underrightarrow{9 \qquad N} 10$

_____

Part O

a. $\underrightarrow{8 \qquad 3} P$

_____

b. $\underrightarrow{6 \qquad W} 8$

_____

c. $\underrightarrow{V \qquad 9} 12$

_____

d. $\underrightarrow{9 \qquad 9} Y$

_____

Remedies

Remedies CONTINUED

Name _____

Part P Write the cents for each row.

a.     [ ] cents

b.   [ ] cents

c.   [ ] cents

Part Q

a. 8 + 10 = _____   8 + 9 = _____

b. 6 + 10 = _____   6 + 9 = _____

c. 1 + 10 = _____   1 + 9 = _____

d. 7 + 10 = _____   7 + 9 = _____

e. 3 + 10 = _____   3 + 9 = _____

Part R

a. 6 + 9 = _____   f. 5 + 9 = _____

b. 8 + 9 = _____   g. 1 + 9 = _____

c. 2 + 9 = _____   h. 3 + 9 = _____

d. 9 + 9 = _____   i. 4 + 9 = _____

e. 7 + 9 = _____

Part S

a. 5 + 2 + 1 = _____   d. 1 + 4 + 3 = _____

b. 9 + 1 + 4 = _____   e. 2 + 2 + 2 = _____

c. 2 + 3 + 1 = _____   f. 7 + 2 + 1 = _____

Part T

a.	b.	c.	d.	e.	f.
4	6	6	2	7	1
2	2	2	2	3	8
+ 1	+ 10	+ 1	+ 3	+ 6	+ 2

Connecting Math Concepts

Mastery Test 3

Name _____

Part 1

a. 6 + 2 = _____

b. 10 – 2 = _____

c. 7 – 5 = _____

d. 2 + 7 = _____

e. 11 – 2 = _____

f. 2 + 3 = _____

g. 6 – 4 = _____

h. 2 + 2 = _____

i. 3 – 1 = _____

j. 9 – 2 = _____

k. 10 – 8 = _____

l. 2 + 5 = _____

m. 6 – 2 = _____

n. 2 + 9 = _____

o. 10 + 2 = _____

Part 2 Write the missing numbers.

a. __70__ __75__ _____ _____ _____ _____

b. __35__ __40__ _____ _____ _____ _____

Part 3

a. 6 + 10 = _____

b. 17 – 7 = _____

c. 19 – 10 = _____

d. 8 + 10 = _____

e. 10 + 10 = _____

f. 2 + 10 = _____

g. 16 – 10 = _____

h. 20 – 10 = _____

i. 4 + 10 = _____

j. 15 – 5 = _____

k. 18 – 8 = _____

l. 3 + 10 = _____

m. 9 + 10 = _____

n. 11 – 10 = _____

o. 12 – 2 = _____

Mastery Test 3   13

Copyright © The McGraw-Hill Companies, Inc. Permission is granted to reproduce for classroom use.

Mastery Test 3

Name _____

Part 4 Write the number of cents.

a.   _____ cents

b.   _____ cents

c.   _____ cents

Part 5 Write the 3 numbers from smallest to biggest.

a. 25   28   23   b. 71   69   72   c. 91   89   88

____ ____ ____   ____ ____ ____   ____ ____ ____

Part 6 Complete the place-value facts.

a. $200 + 30 + 0 =$ _____   b. _____ $= 42$

c. $90 + 7 =$ _____   d. _____ $= 221$

Part 7 Figure out what the letter equals.

a. R is 12 less than 56.

b. 23 is 22 less than P.

c. 57 is 10 more than F.

d. G is 18 more than 41.

Mastery Test 3

Name _____

Part 8 Write the simple place-value fact. Below, write the new place-value fact.

a. _____ + _____ = 92

new _____ + _____ = 92

b. _____ + _____ = 55

new _____ + _____ = 55

Part 9 Write the dollars and cents numbers.

a.

_____ dollars _____ cents

b.

_____ dollars _____ cents

Remedies

Name _____

Part A

a. $4 - 2 =$ ____	i. $6 - 4 =$ ____	q. $5 - 2 =$ ____	x. $10 + 2 =$ ____
b. $8 + 2 =$ ____	j. $2 + 2 =$ ____	r. $9 - 2 =$ ____	y. $3 + 2 =$ ____
c. $11 - 9 =$ ____	k. $10 - 2 =$ ____	s. $7 - 2 =$ ____	z. $10 - 8 =$ ____
d. $5 + 2 =$ ____	l. $12 - 10 =$ ____	t. $3 + 2 =$ ____	A. $7 - 5 =$ ____
e. $5 - 3 =$ ____	m. $9 - 7 =$ ____	u. $11 - 2 =$ ____	B. $6 + 2 =$ ____
f. $12 - 2 =$ ____	n. $6 - 2 =$ ____	v. $9 + 2 =$ ____	C. $11 - 2 =$ ____
g. $8 - 6 =$ ____	o. $4 + 2 =$ ____	w. $8 - 2 =$ ____	D. $11 - 9 =$ ____
h. $7 + 2 =$ ____	p. $7 - 5 =$ ____		

Part B

a. $8 - 2 =$ ____	i. $12 - 10 =$ ____	q. $6 + 2 =$ ____	x. $2 + 8 =$ ____
b. $7 + 2 =$ ____	j. $8 - 6 =$ ____	r. $11 - 9 =$ ____	y. $12 - 2 =$ ____
c. $7 - 5 =$ ____	k. $10 + 2 =$ ____	s. $8 + 2 =$ ____	z. $10 - 2 =$ ____
d. $9 + 2 =$ ____	l. $5 - 3 =$ ____	t. $7 - 2 =$ ____	A. $2 + 2 =$ ____
e. $6 - 4 =$ ____	m. $9 - 7 =$ ____	u. $2 + 9 =$ ____	B. $5 - 2 =$ ____
f. $9 - 2 =$ ____	n. $4 + 2 =$ ____	v. $6 - 2 =$ ____	C. $9 - 2 =$ ____
g. $10 - 8 =$ ____	o. $3 + 2 =$ ____	w. $4 - 2 =$ ____	D. $2 + 7 =$ ____
h. $5 + 2 =$ ____	p. $11 - 2 =$ ____		

Connecting Math Concepts

Remedies

Remedies CONTINUED

Name _____

Part C

a.	b.	c.	d.	e.	f.
6	6	7	7	5	5
+ 10	+ 9	+ 10	+ 9	+ 10	+ 9

Part D

a. $5 + 10 =$ _____

b. $19 - 10 =$ _____

c. $7 + 10 =$ _____

d. $16 - 10 =$ _____

e. $11 - 1 =$ _____

f. $18 - 10 =$ _____

g. $20 - 10 =$ _____

h. $17 - 7 =$ _____

i. $10 + 4 =$ _____

j. $12 - 2 =$ _____

k. $18 - 8 =$ _____

l. $2 + 10 =$ _____

m. $10 + 10 =$ _____

n. $13 - 3 =$ _____

o. $17 - 10 =$ _____

p. $20 - 10 =$ _____

q. $19 - 10 =$ _____

r. $5 + 10 =$ _____

s. $11 - 1 =$ _____

t. $18 - 8 =$ _____

Remedies

Remedies CONTINUED

Name _____

Part E

a. 20 − 10 = _____

b. 7 + 10 = _____

c. 16 − 6 = _____

d. 5 + 10 = _____

e. 14 − 4 = _____

f. 13 − 10 = _____

g. 12 − 2 = _____

h. 18 − 10 = _____

i. 9 + 10 = _____

j. 11 − 10 = _____

k. 15 − 5 = _____

l. 8 + 10 = _____

m. 4 + 10 = _____

n. 12 − 10 = _____

o. 11 − 1 = _____

p. 6 + 10 = _____

q. 10 + 7 = _____

r. 15 − 10 = _____

s. 16 − 10 = _____

t. 10 + 10 = _____

u. 17 − 7 = _____

v. 19 − 10 = _____

w. 18 − 8 = _____

x. 14 − 10 = _____

y. 3 + 10 = _____

z. 17 − 10 = _____

A. 19 − 9 = _____

B. 10 + 6 = _____

C. 1 + 10 = _____

D. 13 − 3 = _____

Part F

a.   cents

b.   cents

c.   cents

Remedies CONTINUED

Name _____

Part G

a.	25	21	20	b.	60	90	70	c.	10	8	12

____ ____ ____   ____ ____ ____   ____ ____ ____

Part H

a.	28	18	24	b.	35	39	36	c.	100	150	50

____ ____ ____   ____ ____ ____   ____ ____ ____

Part I Complete the place-value facts.

a. _____ = 324

b.   600 + 80 + 6 = _____

c. _____ = 609

d.   100 + 0 + 6 = _____

Part J

a. _____ = 63

b.   200 + 0 + 0 = _____

c. _____ = 403

d.   80 + 9 = _____

e. _____ = 781

f. _____ = 571

g.   100 + 10 + 6 = _____

Remedies CONTINUED

Name _____

Part K

a. K is 17 less than 88.

———————→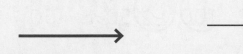

b. 14 is 3 less than F.

———————→

c. M is 22 more than 17.

———————→

d. 34 is 10 more than N.

———————→

e. 26 is 30 less than W.

———————→

Part L

a. N is 15 less than 25.

———————→

b. 26 is 42 less than Y.

———————→

c. T is 22 less than 87.

———————→

d. V is 50 more than 34.

———————→

Part M Complete each equation.

a. ☐ + ☐ = 38

new ☐ + ☐ = 38

b. ☐ + ☐ = 71

new ☐ + ☐ = 71

Connecting Math Concepts

Remedies

Remedies CONTINUED

Name _____

Part N

a. [   ] + [   ] = 23   b. [   ] + [   ] = 76

new [   ] + [   ] = 23   new [   ] + [   ] = 76

Part O

a.

[   ] dollars [   ] cents

b.

[   ] dollars [   ] cents

c.

[   ] dollars [   ] cents

Part P

a.

[   ] dollars [   ] cents

b.

[   ] dollars [   ] cents

c.

[   ] dollars [   ] cents

Mastery Test 4

Name _____

Part 1 Cross out problems you cannot work. Work the other problems.

a. 6
   − 7

b. 6
   − 6

c. 8
   − 9

d. 9
   − 8

e. 10
   − 9

f. 3
   − 5

Part 2 Write the sign >, <, or =.

a. 40 ☐ 39   b. 52 ☐ 52   c. 98 ☐ 100

d. 25 ☐ 27   e. 14 ☐ 12   f. 60 ☐ 60

Part 3 Write **E, S,** and a number for each number family.

a. Ron found 18 coins.

b. Mary gave away 9 cups.

c. Will lost 50 dollars.

d. Jim made 45 cards.

Part 4 Work each problem.

a. 43
   + 12

b. 28
   + 32

c. 59
   + 20

d. 72
   + 19

Part 5 Write the missing numbers for counting by 4s.

___4___ _____ _____ _____ _____

_____ __28__ _____ _____ _____

Mastery Test 4

Name _____

Part 6

a. R is 10 more than P.
R is 60.
What number is P? _____

b. T is 5 less than D.
D is 46.
What number is T? _____

c. N is 11 less than Y.
N is 82.
What number is Y? _____

d. W is 25 more than J.
J is 70.
What number is W? _____

Part 7 Measure each line.

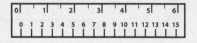

a. ↓_____ ☐ inches

b. ↓_____ ☐ centimeters

Part 8 Write the sign >, <, or =.

a. 20 + 8 ☐ 27   **b.** 45 ☐ 40 + 5

c. 18 − 1 ☐ 18   **d.** 8 − 8 ☐ 1

Mastery Test 4

Name _____

Part 9 Write **E, S**, and a number for each family. Then work the problem.

a. Bob had 29 books.
He lost 19 books.
How many books did he end
up with?

b. The bus started out with 15 people.
32 more people got on the bus.
How many people ended up on
the bus?

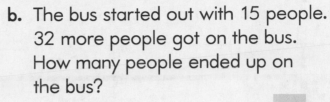

Part 10 Write the missing numbers for counting by 2s.

a. __32__ __34__ ____ ____ ____ ____

b. __56__ __58__ ____ ____ ____ ____

Part 11

a. $25 + 10 =$ _____   b. $406 + 10 =$ _____

c. $10 + 73 =$ _____   d. $10 + 230 =$ _____

Remedies

Name _____

Part A

a.	b.	c.	d.	e.	f.	g.
7	4	6	10	12	8	7
− 8	− 6	− 4	− 10	− 13	− 8	− 6

Part B

a.	b.	c.	d.	e.	f.
7	4	8	7	8	5
− 5	− 3	− 9	− 6	− 8	− 6

Part C

a. 36 ☐ 37   b. 30 ☐ 20   c. 59 ☐ 60   d. 30 ☐ 29

Part D

a. 7 ☐ 6   b. 4 ☐ 4   c. 11 ☐ 12

d. 50 ☐ 60   e. 9 ☐ 8   f. 13 ☐ 13

Part E

a. Fran made 9 more pictures.

b. Bob lost 14 dollars.

Part F

a. Andy sold 17 bikes.

d. They bought 19 boats.

b. Dan gave away 25 dollars.

e. Her mother lost 21 pounds.

c. Tina earned 200 dollars.

Connecting Math Concepts

Remedies

Remedies CONTINUED

Name _____

Part G

a. 38
  + 12

b. 45
  + 39

c. 59
  + 23

Part H

a. 35
  + 23

b. 78
  + 19

c. 45
  + 39

d. 28
  + 31

Part I

___ 8 ___ 16 ___

___ 28 ___ 36 ___

Part J

4 ___ ___ 16 ___

___ ___ 32 ___ ___

Part K

a. T is 18 less than B.
   B is 79.
   What number is T?

   _____

b. R is 54 less than M.
   R is 31.
   What number is M?

   _____

c. W is 22 less than P.
   W is 76.
   What number is P?

   _____

d. Y is 13 more than J.
   Y is 95.
   What number is J?

   _____

   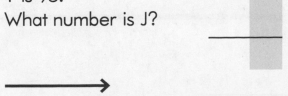

Remedies CONTINUED

Name _____

Part L

a. M is 56 less than J.
M is 30.
What number is J? _____

→

b. R is 20 less than P.
P is 85.
What number is R? _____

→

c. T is 12 more than V.
T is 54.
What number is V? _____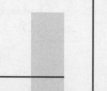

→

d. M is 72 more than F.
F is 23.
What number is M? _____

→

Part M

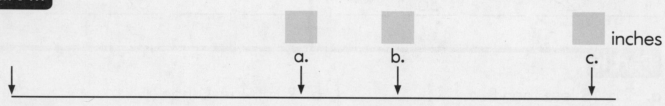

inches

a.   b.   c.

Part N

inches

a.   b.   c.   end

Part O

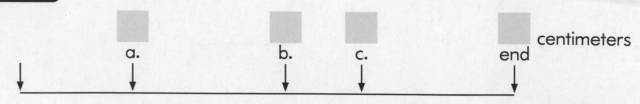

centimeters

a.   b.   c.   end

Connecting Math Concepts

Remedies CONTINUED

Name _____

Part P

a. $5 + 3$ ▢ 10 b. 6 ▢ $7 - 2$

c. $4 + 3$ ▢ 8 d. $12 - 2$ ▢ 9

Part Q

a. $4 - 2$ ▢ 3 d. $9 + 2$ ▢ 11

b. 10 ▢ $7 + 3$ e. $7 - 2$ ▢ 4

c. 14 ▢ $10 + 3$

Part R

a. There were 38 people on the bus. Then 12 people got off the bus. How many people ended up on the bus?

→

b. There were 25 people on a plane. Then 42 people got on the plane. How many people ended up on the plane?

→

c. Tom started with 18 corn chips. He ate 10 chips. How many chips did he still have?

→

Remedies

Remedies CONTINUED

Name _____

Part S

a. Bill had 25 toys.
Then he made 11 toys.
How many toys did Bill end up with?

b. Don had 58 books.
He lost 18 books.
How many books did he end up with?

c. Dolly had 8 bags.
She bought 30 more bags.
How many bags did she end up with?

Part T   Write the missing numbers.

a. __74__ __76__ ____ ____ ____ ____

b. __65__ __70__ ____ ____ ____

Part U

a. 26 + 10 = _____

b. 107 + 10 = _____

c. 10 + 56 = _____

d. 483 + 10 = _____

e. 10 + 103 = _____

f. 10 + 920 = _____

Mastery Test 5

Name _____

Part 1

a. $3 + 5 =$ _____

b. $3 + 8 =$ _____

c. $10 - 3 =$ _____

d. $12 - 3 =$ _____

e. $3 + 4 =$ _____

f. $9 - 3 =$ _____

g. $3 + 2 =$ _____

h. $3 + 7 =$ _____

i. $3 + 10 =$ _____

j. $6 - 3 =$ _____

k. $11 - 3 =$ _____

l. $5 - 3 =$ _____

m. $7 - 3 =$ _____

n. $3 + 9 =$ _____

o. $13 - 3 =$ _____

Part 2 Write the cents for each side. Then make the sign >, <, or =.

a. _____ cents   _____ cents

b. _____ cents   _____ cents

c. _____ cents   _____ cents

Part 3 Write the statement with 3 values on the top line. Below, write the statement about the first and last values.

a. $T < 10$
         $10 < M$

b. $13 > F$
         $F > 10$

_____   _____

_____   _____

Mastery Test 5

Name _____

Part 4 Make a number family. Work the problem.

a. Ron had 18 books.
He bought 12 more books.
How many books did he end up with? ⟶ 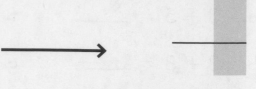 _____

---

b. Beth had some pens.
She lost 4 pens.
She ended up with 25 pens.
How many pens did she start with? ⟶ 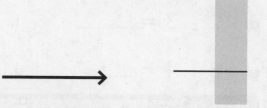 _____

---

c. Sally had some pizza slices.
She made 16 more slices.
She ended up with 26 slices.
How many slices did she start with? ⟶ 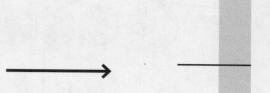 _____

---

d. Bill had 25 dollars.
He spent 14 dollars.
How many dollars did he end up with? ⟶ 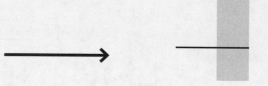 _____

Part 5 Work each problem.

a. 5 x 4 = _____   **b.** 2 x 6 = _____   **c.** 10 x 7 = _____

d. 9 x 2 = _____   **e.** 4 x 3 = _____   **f.** 1 x 5 = _____

Connecting Math Concepts

Mastery Test 5

Name _____

Part 6 Work each problem.

a.	7 1	b.	6 8	c.	4 0	d.	3 5
	− 1 9		− 4 2		− 2 1		− 1 4

Part 7 Measure the line to each arrow.

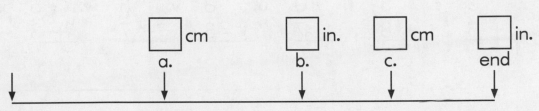

☐ cm   ☐ in.   ☐ cm   ☐ in.
a.   b.   c.   end

Part 8 Complete each equation.

a. 21 + _____ = _____   b. 68 + _____ = _____

Part 9 Write **R**, **T**, or **C** in each shape. Then write **S** in each square.

    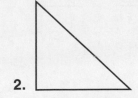

1.   2.   3.   4.

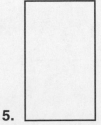

5.   6.   7.   8.

Remedies

Name _____

Part A

a. 3 + 4	b. 6 − 3	c. 9 − 3	d. 8 + 3	e. 7 − 3	f. 3 + 6	g. 10 − 3	h. 7 + 3
i. 3 + 5	j. 7 − 4	k. 8 − 3	l. 3 + 3	m. 11 − 3	n. 5 + 3	o. 9 − 3	p. 3 + 4
q. 11 − 3	r. 3 + 3	s. 3 + 8	t. 10 − 7	u. 8 − 3	v. 3 + 7	w. 6 + 3	x. 6 − 3

Part B

a. 3 + 6	b. 5 + 3	c. 3 + 4	d. 7 − 3	e. 6 − 3	f. 8 + 3	g. 8 − 5	h. 3 + 3
i. 9 − 3	j. 3 + 7	k. 11 − 3	l. 10 − 3	m. 9 + 3	n. 4 − 3	o. 12 − 9	p. 3 + 5
q. 11 − 3	r. 8 − 5	s. 6 + 3	t. 7 − 3	u. 5 + 3	v. 10 − 7	w. 7 − 4	x. 9 + 3

Connecting Math Concepts

Remedies CONTINUED

Name _____

Part C

a. _____ cents    _____ cents

b. _____ cents   _____ cents

c. _____ cents    _____ cents

d. _____ cents    _____ cents

Part D

a. _____ cents    _____ cents

b. _____ cents

Part E

a.   H > C
      C > T

b.   P < J
      J < R

Mastery Test 5 *Remedies* 35

Copyright © The McGraw-Hill Companies, Inc.

Remedies

Remedies CONTINUED

Name _____

Part F

a.
$$8 > K$$
$$K > 3$$

b.
$$7 < B$$
$$B < 9$$

Part G

a. 5 x 4 = _____   b. 2 x 8 = _____

c. 4 x 5 = _____   d. 9 x 3 = _____

Part H

a.
```
  72
- 43
```

b.
```
  72
- 41
```

c.
```
  89
- 17
```

d.
```
  80
- 17
```

e.
```
  61
- 29
```

Part I

a.
```
  62
- 11
```

b.
```
  61
- 12
```

c.
```
  70
- 23
```

d.
```
  92
- 83
```

e.
```
  93
- 82
```

Part J

cm   in.   cm   in.

a.   b.   c.   end

Remedies CONTINUED

Name _____

Part K

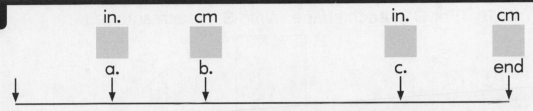

Part L

a. 27

b. 18

c. 73

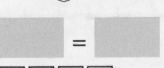

Part M

a. 28

b. 16

c. 39

d. 55

Remedies CONTINUED

Name _____

Part N Write **R**, **T**, or **C** in each shape. Write **S** in each square.

1.
2.
3.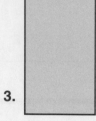
4.
5.

6.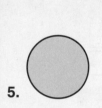
7.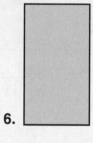
8.
9.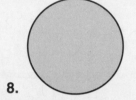

Part O

1.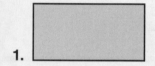
2.
3.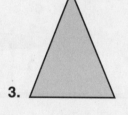
4.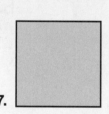

5.
6.
7.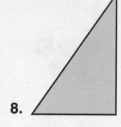
8.
9.

Mastery Test 6

Name _____

Part 1

a. $9 + 8 =$ _____

b. $15 - 6 =$ _____

c. $5 + 9 =$ _____

d. $10 + 6 =$ _____

e. $9 + 9 =$ _____

f. $14 - 5 =$ _____

g. $16 - 7 =$ _____

h. $9 + 3 =$ _____

i. $8 + 9 =$ _____

j. $18 - 8 =$ _____

k. $12 - 3 =$ _____

l. $11 - 9 =$ _____

m. $9 + 4 =$ _____

n. $17 - 8 =$ _____

o. $5 + 10 =$ _____

Part 2

a. Mary had 153 rocks. She found some rocks. She ended up with 195 rocks. How many rocks did she find?

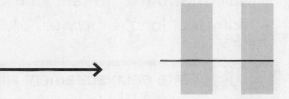

b. Bill started with some dollars. He spent $50. He ended up with $35. How many dollars did he start with?

c. Wendy had 45 apples. She gave away 23 apples. How many apples did she end up with?

Part 3

a. $12 - 6 =$ _____

b. $8 + 8 =$ _____

c. $14 - 7 =$ _____

d. $4 + 4 =$ _____

e. $16 - 8 =$ _____

f. $7 + 7 =$ _____

g. $8 - 4 =$ _____

h. $6 + 6 =$ _____

i. $10 - 5 =$ _____

j. $14 - 7 =$ _____

k. $5 + 5 =$ _____

l. $6 - 3 =$ _____

m. $10 + 10 =$ _____

n. $16 - 8 =$ _____

o. $3 + 3 =$ _____

Mastery Test 6

Name _____

Part 4

a. Bill was 16 inches shorter than Chad. Bill was 43 inches tall. How many inches tall was Chad?

⟶ _____

b. There are 12 fewer red balls than blue balls. There are 65 blue balls. How many red balls are there?

⟶ _____

c. The yard is 25 feet longer than the wall. The yard is 98 feet long. How many feet long is the wall?

⟶ _____

Part 5 Write each statement without the middle value.

a. P > Q
  Q > T

_____

b. 10 < V
  R < 10

_____

c. 6 > M
  11 > 6

_____

Part 6 Complete each equation.

a. 46 + _____ = _____

b. 58 + _____ = _____

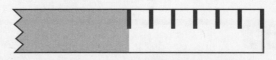

Part 7 Write the letter or letters for each shape: **R, C, T, S, Cu, P, RP, Sp.**

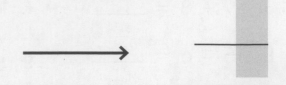 ___
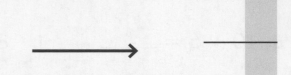 ___
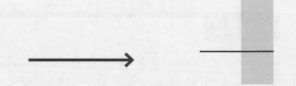 ___
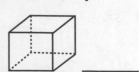 ___

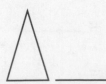 ___
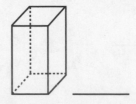 ___
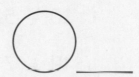 ___
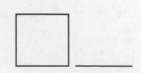 ___

Connecting Math Concepts

Remedies

Name _____

a. 16 b. 6 c. 9 d. 15 e. 14 f. 9 g. 13 h. 16
  − 6   + 9   + 8   − 6   − 5   + 5   − 3   − 7

i. 3 j. 9 k. 9 l. 11 m. 7 n. 13 o. 15 p. 9
  + 9   + 2   + 3   − 1   + 9   − 4   − 5   + 4

q. 5 r. 14 s. 11 t. 17 u. 9 v. 12 w. 8 x. 9
  + 9   − 4   − 2   − 7   + 7   − 2   + 9   + 6

y. 4 z. 17 A. 18 B. 12 C. 18 D. 9
  + 9   − 8   − 9   − 3   − 10   + 9

Remedies CONTINUED

Name _____

Part B

a. 11
− 2

b. 15
− 5

c. 9
+ 5

d. 18
− 9

e. 6
+ 9

f. 13
− 3

g. 16
− 6

h. 9
+ 4

i. 13
− 4

j. 9
+ 8

k. 9
+ 7

l. 11
− 1

m. 7
+ 9

n. 14
− 5

o. 4
+ 9

p. 12
− 3

q. 9
+ 3

r. 16
− 7

s. 14
− 4

t. 3
+ 9

u. 8
+ 9

v. 9
+ 2

w. 12
− 2

x. 15
− 6

y. 9
+ 6

z. 17
− 7

A. 5
+ 9

B. 9
+ 9

C. 17
− 8

D. 18
− 8

Connecting Math Concepts

Remedies CONTINUED

Name _____

a. $3 + 3 =$ _____

b. $8 + 8 =$ _____

c. $10 + 10 =$ _____

d. $7 + 7 =$ _____

e. $5 + 5 =$ _____

f. $9 + 9 =$ _____

g. $2 + 2 =$ _____

h. $6 + 6 =$ _____

i. $1 + 1 =$ _____

j. $4 + 4 =$ _____

k. $6 + 6 =$ _____

l. $5 + 5 =$ _____

m. $1 + 1 =$ _____

n. $7 + 7 =$ _____

o. $3 + 3 =$ _____

p. $10 + 10 =$ _____

q. $9 + 9 =$ _____

r. $4 + 4 =$ _____

s. $8 + 8 =$ _____

t. $2 + 2 =$ _____

Remedies

Remedies CONTINUED

Name _____

Part D

a. 4 + 4 = ____	k. 8 + 8 = ____	u. 10 + 10 = ____
b. 10 + 10 = ____	l. 6 + 6 = ____	v. 14 − 7 = ____
c. 12 − 6 = ____	m. 2 + 2 = ____	w. 5 + 5 = ____
d. 2 + 2 = ____	n. 16 − 8 = ____	x. 6 + 6 = ____
e. 7 + 7 = ____	o. 5 + 5 = ____	y. 7 + 7 = ____
f. 8 + 8 = ____	p. 3 + 3 = ____	z. 8 − 4 = ____
g. 1 + 1 = ____	q. 4 + 4 = ____	A. 3 + 3 = ____
h. 14 − 7 = ____	r. 8 + 8 = ____	B. 1 + 1 = ____
i. 6 + 6 = ____	s. 12 − 6 = ____	C. 9 + 9 = ____
j. 9 + 9 = ____	t. 7 + 7 = ____	D. 4 + 4 = ____

Part E

a. $M < Y$
$Y < 99$

b. $N > F$
$F > P$

_____   _____

Part F

a. $11 < P$
$P < 15$

b. $T < 15$
$F < T$

c. $9 > 6$
$W > 9$

d. $5 > J$
$J > E$

_____ _____   _____ _____

Remedies CONTINUED

Name _____

Part G

a. 41 + ▮ = ▮

b. 26 + ▮ = ▮

Part H

a. 37 + ▮ = ▮

b. 65 + ▮ = ▮

Part I

1. 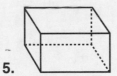   2. □   3. □   4. ▱

5. ▱   6. □   7.   8. ▭

- rectangular prism _____

- square _____

- cube _____

- rectangle _____

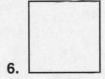

Remedies CONTINUED

Name _____

1.

2.

3.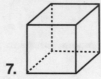

4.

5.

6.

7.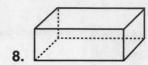

8.

• cube _____

• rectangle _____

• rectangular prism _____

• square _____

Mastery Test 7

Name _____

Part 1 Write each answer.

a. 62 + 9 = _____   b. 17 + 9 = _____   c. 43 + 9 = _____

d. 88 + 9 = _____   e. 31 + 9 = _____   f. 15 + 9 = _____

Part 2 Write each dollars and cents number: $■■.■■

a.

 _____

b.

 _____

Part 3 Write each answer.

a. 14 – 7 = _____   f. 6 + 6 = _____   k. 7 + 9 = _____

b. 8 – 6 = _____   g. 7 – 3 = _____   l. 17 – 7 = _____

c. 8 + 8 = _____   h. 8 + 10 = _____   m. 10 – 5 = _____

d. 15 – 10 = _____   i. 15 – 6 = _____   n. 11 – 8 = _____

e. 4 + 9 = _____   j. 7 – 4 = _____   o. 10 + 10 = _____

Mastery Test 7

Name _____

a. Wendy earned $188. Bill earned $199. How many more dollars did Bill earn than Wendy?

b. Fran is 6 pounds lighter than Todd. Fran weighs 43 pounds. How many pounds does Todd weigh?

c. There are 28 sheep and 18 cows on the farm. How many fewer cows than sheep are there?

Part 5 Complete each equation.

a. _____ + _____ = 79

b. 68 + _____ = _____

Copyright © The McGraw-Hill Companies, Inc. Permission is granted to reproduce for classroom use.

Connecting Math Concepts

Mastery Test 7

Name _____

Part 6 For each family, figure out what **t** equals.

a. $\overset{r}{\underset{}{t \longrightarrow}} d$

d = 58
r = 27

b. $\overset{k}{\underset{}{m \longrightarrow}} t$

k = 15
m = 45

c. $\overset{t}{\underset{}{v \longrightarrow}} j$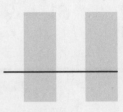

v = 417
j = 579

Part 7 Figure out each missing number.

a. 10 + ____ = 16   b. 5 + ____ = 10   c. 3 + ____ = 7

d. 20 + ____ = 28   e. 2 + ____ = 9   f. 6 + ____ = 12

Remedies

Name _____

Part A

a. $46 + 10 =$ _____

$46 + 9 =$ _____

b. $58 + 10 =$ _____

$58 + 9 =$ _____

c. $32 + 10 =$ _____

$32 + 9 =$ _____

d. $88 + 10 =$ _____

$88 + 9 =$ _____

e. $61 + 10 =$ _____

$61 + 9 =$ _____

f. $25 + 10 =$ _____

$25 + 9 =$ _____

Part B

a. $52 + 9 =$ _____

b. $35 + 9 =$ _____

c. $67 + 9 =$ _____

d. $11 + 9 =$ _____

e. $48 + 9 =$ _____

f. $39 + 9 =$ _____

Part C

a. $\begin{array}{r} 14 \\ -7 \\ \hline \end{array}$
b. $\begin{array}{r} 18 \\ -10 \\ \hline \end{array}$
c. $\begin{array}{r} 17 \\ -8 \\ \hline \end{array}$
d. $\begin{array}{r} 15 \\ -6 \\ \hline \end{array}$
e. $\begin{array}{r} 7 \\ +9 \\ \hline \end{array}$
f. $\begin{array}{r} 4 \\ +9 \\ \hline \end{array}$
g. $\begin{array}{r} 14 \\ -5 \\ \hline \end{array}$
h. $\begin{array}{r} 10 \\ -3 \\ \hline \end{array}$

i. $\begin{array}{r} 8 \\ +9 \\ \hline \end{array}$
j. $\begin{array}{r} 6 \\ +9 \\ \hline \end{array}$
k. $\begin{array}{r} 8 \\ -3 \\ \hline \end{array}$
l. $\begin{array}{r} 12 \\ -3 \\ \hline \end{array}$
m. $\begin{array}{r} 7 \\ -4 \\ \hline \end{array}$
n. $\begin{array}{r} 8 \\ -5 \\ \hline \end{array}$
o. $\begin{array}{r} 11 \\ -3 \\ \hline \end{array}$

Part D

a. $10 - 3 =$ _____

e. $5 + 9 =$ _____

i. $15 - 6 =$ _____

m. $7 - 5 =$ _____

b. $7 - 4 =$ _____

f. $9 - 6 =$ _____

j. $13 - 3 =$ _____

n. $18 - 10 =$ _____

c. $9 + 9 =$ _____

g. $14 - 7 =$ _____

k. $7 + 9 =$ _____

o. $18 - 8 =$ _____

d. $16 - 8 =$ _____

h. $8 + 9 =$ _____

l. $3 + 9 =$ _____

p. $5 - 5 =$ _____

Connecting Math Concepts

Remedies CONTINUED

Name _____

Part E

a. ▢ + ▢ = 61

b. ▢ + ▢ = 19

Part F

a. 53 + ▢ = ▢

b. ▢ + ▢ = 81

Part G

a. 10 + ■ = 14

b. 9 + ■ = 12

c. 3 + ■ = 5

⟶

⟶

⟶

d. 2 + ■ = 11

e. 1 + ■ = 8

⟶

⟶

Part H

a. 3 + ▢ = 10

b. 8 + ▢ = 16

c. 10 + ▢ = 20

d. 4 + ▢ = 7

e. 2 + ▢ = 12

Remedies

Cumulative Test 1

(After Lesson 70)

Name _____

Part 1

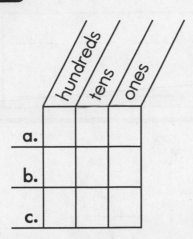

	hundreds	tens	ones
a.			
b.			
c.			

Part 2 Write 4 facts for the family.

a. $\underrightarrow{\quad 4 \qquad 2 \quad} 6$

_____

_____

_____

_____

Part 3

a. 612

hundreds digit _____

ones digit _____

b. 40

ones digit _____

tens digit _____

Part 4 Write the fact. Then write the missing number.

a. $\underrightarrow{\quad 7 \qquad 2 \quad}$ ___

_____

b. $\underrightarrow{\quad 9 \qquad} 10$

_____

Cumulative Test 1

Name _____

Part 5 Write the missing numbers.

a. _5_ _10_ ___ ___ ___ _30_ ___ ___ ___ ___

b. _2_ ___ ___ ___ _10_ ___ ___ ___ _18_ ___

c. _9_

___

___

_36_

___

___

___

___

_81_

___

Part 6

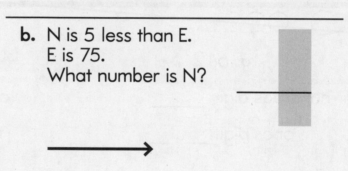

a. M is 10 more than T.
M is 80.
What number is T?

_____

b. N is 5 less than E.
E is 75.
What number is N?

_____

Part 7 Measure the line to each arrow.

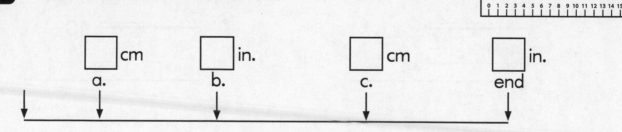

☐ cm   ☐ in.   ☐ cm   ☐ in.
a.   b.   c.   end

Part 8 Write the 3 numbers from smallest to biggest.

a. 51   49   53   b. 24   27   22

___ ___ ___   ___ ___ ___

Connecting Math Concepts

Cumulative Test 1

Name _____

Part 9 Write the sign >, <, or =.

a. 28 ☐ 27

b. 32 ☐ 30 + 2

c. 12 ☐ 10 + 1

d. 15 ☐ 20

Part 10 Complete each equation.

a.   65 + _____ = _____

b.   68 + _____ = _____

Part 11 Write the statement with 3 values on the top line. Below, write the statement about the first and last values.

a.   28 > F

            F > 25

b.   P < 60

           60 < W

Cumulative Test 1 Name _____

Part 12

a. Pat had 145 pens. She found some pens. She ended up with 158 pens. How many pens did she find? ⟶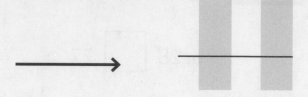

b. Sally started with some dollars. She spent $40. She ended up with $25. How many dollars did she start with? ⟶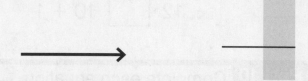

Part 13 Work each problem.

a. 5 x 6 = _____   b. 2 x 4 = _____

Part 14 Write the letter or letters for each shape: **R, C, T, S, Cu, P, RP, Sp.**

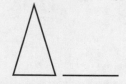 _____    _____   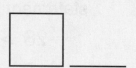 _____   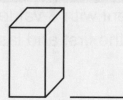 _____

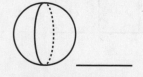 _____   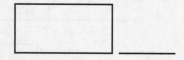 _____   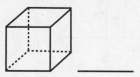 _____   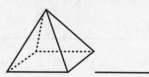 _____

Part 15 Write each answer.

a. 10 + 45 = _____   b. 308 + 10 = _____

Part 16 Figure out each missing number.

a. 10 + _____ = 19   b. 5 + _____ = 10

Cumulative Test 1

Name _____

Part 17 Work each problem.

a.
```
   6
   1
 + 3
─────
```

b.
```
   7
   3
 + 4
─────
```

Part 18 Write the missing numbers for counting by 2s.

a. <u>52</u> <u>54</u> ___ ___ ___ ___

Part 19 Work each problem.

a.
```
  8 1
 -1 9
──────
```

b.
```
  1 8
 +4 2
──────
```

c.
```
  8 7
 -2 6
──────
```

d.
```
  3 9
 +1 5
──────
```

Part 20 For each family, figure out what **v** equals.

a.
$$\underrightarrow{\text{v} \qquad \text{z}} \text{n}$$
n = 48
z = 21

b.
$$\underrightarrow{\text{h} \qquad \text{f}} \text{v}$$
f = 18
h = 42

Part 21 Write the missing numbers for counting by 4s.

<u>4</u> ___ ___ ___ ___

___ <u>28</u> ___ ___ ___

Part 22 Complete the place-value facts.

a. 500 + 70 + 0 = _____

b. 40 + 6 = _____

Mastery Test 8

Name _____

Part 1 Write the answer.

a. $14 - 10 =$ _____ f. $13 - 9 =$ _____ k. $9 - 5 =$ _____

b. $9 + 4 =$ _____ g. $4 + 10 =$ _____ l. $14 - 4 =$ _____

c. $4 + 6 =$ _____ h. $9 - 4 =$ _____ m. $4 + 9 =$ _____

d. $8 - 4 =$ _____ i. $5 + 4 =$ _____ n. $6 + 4 =$ _____

e. $4 + 5 =$ _____ j. $10 - 6 =$ _____ o. $13 - 4 =$ _____

Part 2 Make a number family with 3 letters. Find the missing number.

a. 19 people
11 women
How many men?

b. 15 dogs
How many pets?
23 cats

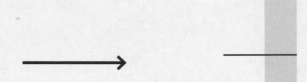

c. How many plums?
40 apples
68 fruits

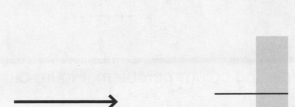

Part 3 Work each problem.

a.
```
  1 3 8
+   4 2
```

b.
```
  5 7 1
- 1 1 2
```

c.
```
  4 5 9
+ 1 2 7
```

d.
```
  6 5 0
-   4 5
```

Mastery Test 8

Name _____

Part 4 Write the closer tens number.

a. 74 _____   b. 39 _____   c. 56 _____   d. 23 _____

Part 5 Write the answer.

a. 80 + 80 = _____   b. 110 – 30 = _____   c. 90 – 80 = _____

d. 40 + 20 = _____   e. 140 – 70 = _____   f. 50 + 50 = _____

Part 6 Find the perimeter of each figure. Remember the unit name.

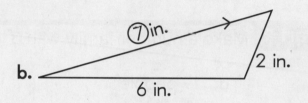

Part 7 Write the column problem. Figure out the missing number.

a. 146 – _____ = 104   b. 73 + _____ = 91   c. _____ + 150 = 261

Remedies

Name _____

Part A

a. $5 + 4 =$ ___	f. $14 - 10 =$ ___	k. $9 - 5 =$ ___	p. $13 - 9 =$ ___
b. $10 - 4 =$ ___	g. $8 - 4 =$ ___	l. $4 + 9 =$ ___	q. $10 - 6 =$ ___
c. $4 + 4 =$ ___	h. $4 + 6 =$ ___	m. $14 - 4 =$ ___	r. $4 + 10 =$ ___
d. $10 + 4 =$ ___	i. $4 + 5 =$ ___	n. $10 - 6 =$ ___	s. $9 - 4 =$ ___
e. $13 - 4 =$ ___	j. $9 + 4 =$ ___	o. $6 + 4 =$ ___	t. $5 + 4 =$ ___

Part B

a. $14 - 10 =$ ___	f. $13 - 9 =$ ___	k. $5 + 4 =$ ___	p. $4 + 9 =$ ___
b. $4 + 6 =$ ___	g. $4 + 10 =$ ___	l. $10 - 4 =$ ___	q. $14 - 4 =$ ___
c. $8 - 4 =$ ___	h. $10 - 6 =$ ___	m. $4 + 4 =$ ___	r. $10 - 6 =$ ___
d. $9 + 4 =$ ___	i. $9 - 4 =$ ___	n. $13 - 4 =$ ___	s. $9 - 5 =$ ___
e. $4 + 5 =$ ___	j. $6 + 4 =$ ___	o. $10 + 4 =$ ___	t. $6 + 4 =$ ___

Part C

a.
$$\begin{array}{r} 176 \\ +\ 316 \\ \hline \end{array}$$

b.
$$\begin{array}{r} 265 \\ -\ 129 \\ \hline \end{array}$$

c.
$$\begin{array}{r} 346 \\ -\ 237 \\ \hline \end{array}$$

Part D

a. $160 - 150 =$ ___   b. $110 - 80 =$ ___   c. $120 - 100 =$ ___

Remedies CONTINUED

Name _____

Part E

a. 120 – 10 = _____   **b.** 160 – 10 = _____   **c.** 160 – 80 = _____

d. 90 + 40 = _____   **e.** 120 + 20 = _____   **f.** 10 + 170 = _____

Part F

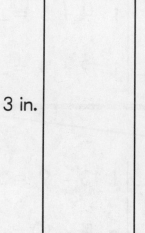

b.

3 in.

1 in.

a.

6 cm

1 cm

_____ =

_____ =

_____ =

_____ =

Remedies CONTINUED

Name _____

Part G

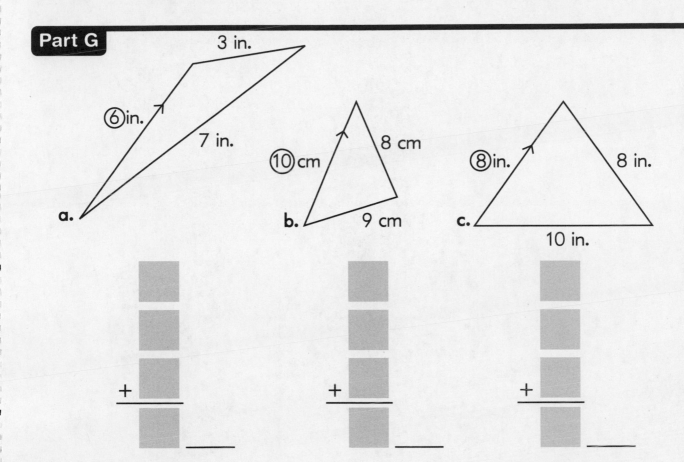

Mastery Test 9

Name _____

Part 1 Write the unit name for each question.

a. How many new pens are there? 15 _____

b. How many inches taller is Sally? 8 _____

c. How many full cups are there? 35 _____

d. How many trucks are in the lot? 125 _____

Part 2 Write the answer.

a. $12 - 7 =$ _____   **f.** $12 - 5 =$ _____   **k.** $12 - 7 =$ _____

b. $5 + 6 =$ _____   **g.** $5 + 6 =$ _____   **l.** $6 + 5 =$ _____

c. $11 - 6 =$ _____   **h.** $11 - 6 =$ _____   **m.** $12 - 5 =$ _____

d. $7 + 5 =$ _____   **i.** $5 + 7 =$ _____   **n.** $7 + 5 =$ _____

e. $6 + 5 =$ _____   **j.** $11 - 5 =$ _____   **o.** $11 - 5 =$ _____

Part 3 Work each problem. Show the answer with a unit name.

a. 16 children and 83 adults are at the show. How many people are at the show?

b. 65 cars are on the lot. 12 of the cars are new. How many used cars are on the lot?

c. There are 310 boys at school. If there are 623 children at school, how many girls are there?

Mastery Test 9

Name _____

Mastery Test 9

Part 4 Work the times problem for each row.

a.

b.

c.

Part 5

a. Rob started with 68 dollars.
He spent some dollars.
He ended up with 27 dollars.
How many dollars did he spend?

_____

b. Ana had 250 stamps.
She bought 29 more stamps.
How many stamps did she end up with?

_____

c. 15 ripe apples fell from the tree.
The tree ended up with 74 ripe apples.
How many ripe apples did the tree
start with?

Part 6 Work each problem.

$$
\begin{array}{r} a.\quad 369 \\ +172 \\ \hline \end{array}
\qquad
\begin{array}{r} b.\quad 163 \\ -\ 81 \\ \hline \end{array}
\qquad
\begin{array}{r} c.\quad 407 \\ -155 \\ \hline \end{array}
$$

Mastery Test 9

Name _____

Part 7 Write each estimation problem and work it.

a.
```
   6 2
 - 3 8
```

b.
```
   5 8
 + 2 9
```

c.
```
   9 4
 - 7 2
```

Part 8 Write the time for each clock.

a.

b.

c.

Part 9 Write each problem in a column and work it.

a. 200 + 47 + 7 + 120

b. 103 + 9 + 50

Remedies

Name _____

Part A

a. How many black cats did they have?   17 _____

b. How many small boys were in the room?   32 _____

c. How many big boxes are on the truck?   108 _____

d. How many long snakes were there?   5 _____

e. How many blue birds are in the tree?   21 _____

Part B

a. How many short pencils are there?   10 _____

b. How many inches shorter is the fence?   216 _____

c. How many white goats does Sally have?   25 _____

d. How many broken eggs were there?   144 _____

e. How many cars are in the lot?   93 _____

Part C

a. $6 + 5 =$ _____   **f.** $15 - 10 =$ _____   **k.** $11 - 6 =$ _____   **p.** $14 - 5 =$ _____

b. $12 - 5 =$ _____   **g.** $10 - 5 =$ _____   **l.** $5 + 9 =$ _____   **q.** $12 - 7 =$ _____

c. $5 + 5 =$ _____   **h.** $5 + 7 =$ _____   **m.** $15 - 5 =$ _____   **r.** $5 + 10 =$ _____

d. $10 + 5 =$ _____   **i.** $5 + 6 =$ _____   **n.** $12 - 7 =$ _____   **s.** $11 - 5 =$ _____

e. $14 - 5 =$ _____   **j.** $9 + 5 =$ _____   **o.** $5 + 7 =$ _____   **t.** $6 + 5 =$ _____

Remedies CONTINUED

Name _____

Part D

a. 15 − 10 = _____   f. 14 − 9 = _____   k. 6 + 5 = _____   p. 5 + 9 = _____

b. 5 + 7 = _____   g. 5 + 10 = _____   l. 12 − 5 = _____   q. 15 − 5 = _____

c. 10 − 5 = _____   h. 12 − 7 = _____   m. 5 + 5 = _____   r. 12 − 7 = _____

d. 9 + 5 = _____   i. 11 − 6 = _____   n. 14 − 5 = _____   s. 11 − 5 = _____

e. 5 + 6 = _____   j. 7 + 5 = _____   o. 10 + 5 = _____   t.   7 + 5 = _____

Part E

a.   368   b.   159   c.   159   d.   792
   + 213   + 450   + 415   + 83

Part F

   a.   408   b.   769   c.   517
      − 122   − 81   − 293

Part G

a.   813   b.   629   c.   457   d.   590
   − 281   + 194   + 93   − 45

Part H

a.   36      b.   81      c.   17   ⬜
   − 12   − 48   + 41

Remedies

Mastery Test 10   Name _____

Part 1

a. $7 + 6 =$ _____

b. $14 - 8 =$ _____

c. $14 - 6 =$ _____

d. $6 + 8 =$ _____

e. $13 - 7 =$ _____

f. $14 - 6 =$ _____

g. $6 + 7 =$ _____

h. $13 - 6 =$ _____

i. $7 + 6 =$ _____

j. $8 + 6 =$ _____

k. $13 - 7 =$ _____

l. $8 + 6 =$ _____

m. $14 - 8 =$ _____

n. $6 + 7 =$ _____

o. $6 + 8 =$ _____

Part 2   Work a times problem for each item. Then write the unit name.

a. How many cents is 5 dimes?

b. How many cents is 3 quarters?

c. How many cents is 8 nickels?

Part 3   Check each answer. If the answer is wrong, cross it out and fix it.

a.	b.	c.	d.	e.
5	5	3	1	4
2	5	7	8	5
+ 7	+ 5	+ 2	+ 9	+ 5
16	15	11	19	14

Part 4   Write the sign >, <, or =.

a. 1 yard ☐ 2 feet

b. 23 hours ☐ 1 day

c. 7 days ☐ 1 week

d. 105 cents ☐ 1 dollar

e. 1 hour ☐ 58 minutes

f. 1 gallon ☐ 4 quarts

g. 11 inches ☐ 1 foot

h. 1 year ☐ 13 months

Mastery Test 10

Name _____

Part 5 | Work each problem.

a. $2 \times \underline{\hspace{1cm}} = 10$ **b.** $4 \times \underline{\hspace{1cm}} = 28$ **c.** $9 \times \underline{\hspace{1cm}} = 36$

d. $10 \times \underline{\hspace{1cm}} = 30$ **e.** $5 \times \underline{\hspace{1cm}} = 25$ **f.** $4 \times \underline{\hspace{1cm}} = 24$

Part 6 | Write the time for each clock.

a.

b.

c.

d.

Part 7 | Work each problem.

a.
$$\begin{array}{r} \$79.39 \\ -6.49 \\ \hline \end{array}$$

b.
$$\begin{array}{r} \$1.52 \\ +110.45 \\ \hline \end{array}$$

c.
$$\begin{array}{r} \$.25 \\ 7.50 \\ +.15 \\ \hline \end{array}$$

Remedies

Name _____

Part A

a. $7 + 6 =$ _____	f. $16 - 10 =$ _____	k. $13 - 7 =$ _____	p. $15 - 6 =$ _____
b. $13 - 6 =$ _____	g. $12 - 6 =$ _____	l. $6 + 9 =$ _____	q. $14 - 8 =$ _____
c. $6 + 6 =$ _____	h. $8 + 6 =$ _____	m. $16 - 6 =$ _____	r. $6 + 10 =$ _____
d. $10 + 6 =$ _____	i. $6 + 7 =$ _____	n. $14 - 8 =$ _____	s. $14 - 6 =$ _____
e. $15 - 9 =$ _____	j. $9 + 6 =$ _____	o. $6 + 8 =$ _____	t. $7 + 6 =$ _____

Part B

a.	b.	c.	d.
8	8	2	10
1	2	1	5
+ 9	+ 6	+ 8	+ 5
18	15	12	20

Part C

a.	b.	c.	d.
8	8	3	5
7	8	1	3
+ 3	+ 1	+ 4	+ 6
18	16	9	14

Part D

a. 1 gallon ☐ 5 quarts   e. 9 days ☐ 1 week

b. 1 hour ☐ 60 minutes   f. 1 day ☐ 24 hours

c. 10 inches ☐ 1 foot   g. 14 months ☐ 1 year

d. 1 foot ☐ 1 yard   h. 1 dollar ☐ 99 cents

Remedies

Remedies CONTINUED

Name _____

Part E

a. 1 week ▢ 6 days

b. 4 quarts ▢ 1 gallon

c. 1 year ▢ 18 months

d. 95 cents ▢ 1 dollar

e. 1 foot ▢ 12 inches

f. 1 yard ▢ 2 feet

g. 22 hours ▢ 1 day

h. 70 minutes ▢ 1 hour

Part F

a. $5 \times \underline{\hspace{1cm}} = 40$ b. $9 \times \underline{\hspace{1cm}} = 45$ c. $10 \times \underline{\hspace{1cm}} = 40$ d. $4 \times \underline{\hspace{1cm}} = 28$

Part G

a. $2 \times \underline{\hspace{1cm}} = 12$ b. $5 \times \underline{\hspace{1cm}} = 20$ c. $9 \times \underline{\hspace{1cm}} = 36$ d. $4 \times \underline{\hspace{1cm}} = 32$

Part H

a.
$$\begin{array}{r} \$7.52 \\ -3.19 \\ \hline \end{array}$$

b.
$$\begin{array}{r} \$4.49 \\ +\ .15 \\ \hline \end{array}$$

c.
$$\begin{array}{r} \$3.85 \\ -1.90 \\ \hline \end{array}$$

d.
$$\begin{array}{r} \$5.08 \\ -4.25 \\ \hline \end{array}$$

Connecting Math Concepts

Remedies

Mastery Test 11   Name _____

a. 13 – 8 = _____   f. 12 – 8 = _____   k. 12 – 4 = _____

b. 8 + 4 = _____   g. 8 + 5 = _____   l. 3 + 8 = _____

c. 8 + 3 = _____   h. 11 – 3 = _____   m. 13 – 5 = _____

d. 5 + 8 = _____   i. 4 + 8 = _____   n. 12 – 8 = _____

e. 11 – 8 = _____   j. 13 – 8 = _____   o. 5 + 8 = _____

Part 2 Work each item.

$2.50

$1.50

$14.49

$3.40

$15.75

a. You buy the cake. How much money do you still have?

b. You buy the ice cream. How much money do you still have?

c. You buy the cake and the balloons. Do you have enough money?

yes / no

d. You buy the cake and the candles. Do you have enough money?

yes / no

Mastery Test 11

Name _____

Part 3

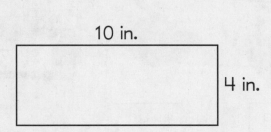

	chairs	tables	books
room T	11	25	5
room W	2	12	10
room Y	5	30	0

a. 11 _____ in room _____

b. 10 _____ in room _____

c. 30 _____ in room _____

Part 4 Find the perimeter and the area of each rectangle.

10 in.

4 in.

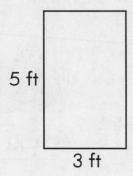

5 ft

3 ft

a. Perimeter

c. Perimeter

b. Area

d. Area

Connecting Math Concepts

Mastery Test 11

Name _____

Part 5 Work each item.

a. Maria has $3.50 less than Pablo. Maria has $10.18. How much does Pablo have?

b. Pedro has $10.45. Tomás has $20.75. How much more does Tomás have than Pedro?

Part 6 Write each answer.

a. $150 + 20 =$ _____

b. $630 + 10 =$ _____

c. $210 + 40 =$ _____

d. $120 + 30 =$ _____

Part 7 Work each item.

a. Write 329 cents with a $ sign. _____

b. Write 500 cents with a $ sign. _____

c. Write 8 cents with a $ sign. _____

d. Write 60 cents with a $ sign. _____

Mastery Test 11

Name _____

Part 8 Work the column problem for each item.

a. _____ − 26 = 54

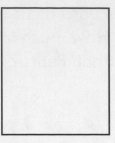

c. _____ + 35 = 38

b. 60 − _____ = 28

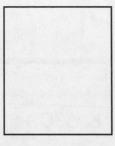

d. _____ − 11 = 43

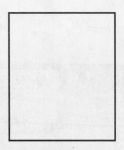

Part 9 Figure out the number of rows or squares.

a.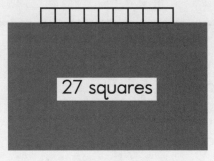

27 squares

_____

b.

6 rows

_____

Part 10 Write the sign >, <, or =.

a. 4 feet ☐ 1 yard

b. 4 quarts ☐ 1 gallon

c. 1 meter ☐ 110 centimeters

d. 1 hour ☐ 60 minutes

e. 22 hours ☐ 1 day

f. 1 foot ☐ 11 inches

Remedies

Name _____

Part A

a. 11 − 8 = _____   f. 8 + 4 = _____   k. 8 + 5 = _____

b. 8 + 2 = _____   g. 8 + 6 = _____   l. 8 + 1 = _____

c. 8 + 5 = _____   h. 14 − 6 = _____   m. 12 − 8 = _____

d. 13 − 8 = _____   i. 8 + 4 = _____   n. 13 − 8 = _____

e. 12 − 8 = _____   j. 10 − 8 = _____   o. 8 + 3 = _____

Part B

	lions	tigers	bears
zoo X	12	6	5
zoo Y	3	10	2
zoo Z	8	4	0

a. 12 _____ in zoo _____

b. 4 _____ in zoo _____

c. 2 _____ in zoo _____

d. 10 _____ in zoo _____

Part C

	dogs	cats	birds
barn	6	12	18
home	2	3	1
woods	4	5	80

a. 3 _____ in the _____

b. 18 _____ in the _____

c. 2 _____ in the _____

d. 80 _____ in the _____

Part D

a. 170 + 20 = _____   b. 320 − 20 = _____

c. 130 + 60 = _____   d. 30 + 120 = _____

Remedies CONTINUED

Name _____

Part E

a. 437 cents = ▮▮▮▮

b. 608 cents = ▮▮▮▮

c. 410 cents = ▮▮▮▮

d. 182 cents = ▮▮▮▮

Part F

a. _____ $- 7 = 3$ _____

b. _____ $- 2 = 9$ _____

c. _____ $+ 5 = 10$ _____

d. _____ $- 5 = 10$ _____

e. _____ $+ 6 = 10$ _____

Part G

a. _____ $- 3 = 9$ _____

b. _____ $+ 2 = 10$ _____

c. _____ $- 7 = 9$ _____

d. _____ $- 9 = 10$ _____

e. _____ $+ 6 = 12$ _____

Part H

a. 25 seconds ▮ 1 minute

b. 1 meter ▮ 65 centimeters

c. 1 week ▮ 9 days

d. 12 months ▮ 1 year

e. 110 centimeters ▮ 1 meter

f. 1 yard ▮ 3 feet

g. 1 gallon ▮ 3 quarts

h. 95 cents ▮ 1 dollar

Part I

a. 1 year ▮ 11 months

b. 75 seconds ▮ 1 minute

c. 1 day ▮ 28 hours

d. 7 days ▮ 1 week

e. 1 meter ▮ 110 centimeters

f. 11 feet ▮ 1 yard

g. 1 minute ▮ 60 seconds

h. 1 quart ▮ 1 gallon

Connecting Math Concepts

Mastery Test 12

Name _____

Part 1

a. $4 + 6 =$ _____

b. $11 - 4 =$ _____

c. $7 + 6 =$ _____

d. $15 - 7 =$ _____

e. $12 - 4 =$ _____

f. $8 + 6 =$ _____

g. $13 - 6 =$ _____

h. $8 + 7 =$ _____

i. $8 + 5 =$ _____

j. $13 - 5 =$ _____

k. $7 + 4 =$ _____

l. $13 - 8 =$ _____

m. $10 - 6 =$ _____

n. $5 + 6 =$ _____

o. $8 + 4 =$ _____

Part 2   Work each item.

$15.50 $12.25 $11.55 $10.25 $59.85

a. You buy the gloves and the scarf. How much money do you still have?

b. You buy the hat, the socks, and the gloves. How much money do you still have?

Part 3   Answer each question.

	hats	shirts	pants
green	3	11	0
white	15	8	10
red	21	5	16

a. How many white pants are there? _____

b. The most hats are which color? _____

c. There are 11 green _____.

d. How many shirts are white? _____

Mastery Test 12

Name _____

Part 4 Find the total length. Find the difference.

cm

a. _____↓

cm

b. _____↓

c. big line: _____

d. difference: _____

Part 5 Answer each question. Remember the unit name.

☐

a.

☐

• The top row shows how many cups are on each shelf.

How many cups are on 6 shelves?

☐

_____

☐

b.

☐

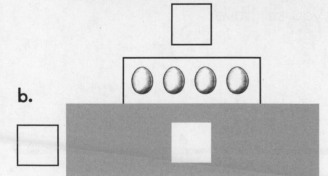

• The top row shows how many eggs are in each bag.

How many bags are there for 20 eggs?

☐

_____

Part 6 Write the sign >, <, or =.

a. 240 ☐ 204

b. 749 ☐ 700 + 50 + 1

c. 400 + 90 + 9 ☐ 501

d. 116 ☐ 115

---

82 Mastery Test 12

Connecting Math Concepts

Remedies

Name _____

Part A

a. 5 + 7 = _____	f. 8 + 6 = _____	k. 4 + 8 = _____	p. 6 + 5 = _____
b. 8 + 4 = _____	g. 4 + 6 = _____	l. 8 + 5 = _____	q. 6 + 8 = _____
c. 5 + 4 = _____	h. 5 + 8 = _____	m. 7 + 5 = _____	r. 7 + 4 = _____
d. 5 + 6 = _____	i. 4 + 7 = _____	n. 6 + 4 = _____	s. 8 + 7 = _____
e. 6 + 7 = _____	j. 7 + 8 = _____	o. 4 + 5 = _____	t. 7 + 6 = _____

Part B

a. 8 + 5 = _____	f. 11 − 4 = _____	k. 8 + 4 = _____	p. 7 + 4 = _____
b. 14 − 6 = _____	g. 4 + 6 = _____	l. 13 − 5 = _____	q. 12 − 5 = _____
c. 7 + 5 = _____	h. 15 − 7 = _____	m. 6 + 5 = _____	r. 6 + 8 = _____
d. 11 − 5 = _____	i. 12 − 4 = _____	n. 8 + 7 = _____	s. 10 − 4 = _____
e. 9 − 5 = _____	j. 7 + 6 = _____	o. 13 − 6 = _____	t. 5 + 4 = _____

Part C

a. _____→ cm

b. _____→ cm

big line: _____

difference: _____

Remedies CONTINUED

Name _____

Part D

cm

a. _____↓

cm

b. _____↓

big line: _____

difference: _____

Part E

cm

a. _____↓

cm

b. _____↓

big line: _____

difference: _____

Part F

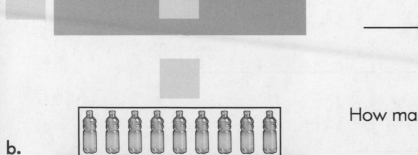

a.   How many rows are there for 20 chairs?

_____

b.   How many bottles are in 10 rows?

_____

Remedies CONTINUED

Name _____

Part G

a.

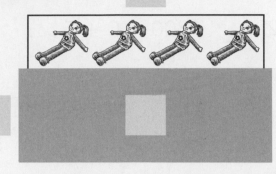

How many boxes are there for 40 toys?

b.

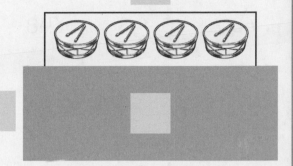

How many toys are in 10 boxes?

c.

How many boats are there for 30 children?

_____

Part H

a. 301 ⬜ 310

b. 518 ⬜ 515

c. 625 ⬜ 725

d. 620 ⬜ 602

e. 110 ⬜ 109

f. 123 ⬜ 210

Remedies

Remedies CONTINUED

Name _____

Part I

a. 600 + 40 + 3 ☐ 648

b. 200 + 0 + 6 ☐ 260

c. 151 ☐ 100 + 10 + 5

d. 225 ☐ 300 + 0 + 5

e. 100 + 80 + 0 ☐ 179

Part J

a. 395 ☐ 401

b. 648 ☐ 651

c. 830 ☐ 803

d. 145 ☐ 205

e. 790 ☐ 709

f. 301 ☐ 299

Part K

a. 148 ☐ 100 + 80 + 4

b. 600 + 70 + 8 ☐ 765

c. 710 ☐ 700 + 0 + 9

d. 515 ☐ 500 + 10 + 8

e. 300 + 20 + 0 ☐ 420

Remedies

Mastery Test 13   Name _____

a. Margo started with some eggs.
She used 9 eggs. She ended up
with 12 eggs. How many eggs did
she start with?

_____

b. Isabel is 39 years older than
Cristi. Isabel is 58 years old.
How old is Cristi?

_____

c. There are 16 apples and
10 bananas on a plate. How
many pieces of fruit are on
the plate?

Part 2 Make a line plot.

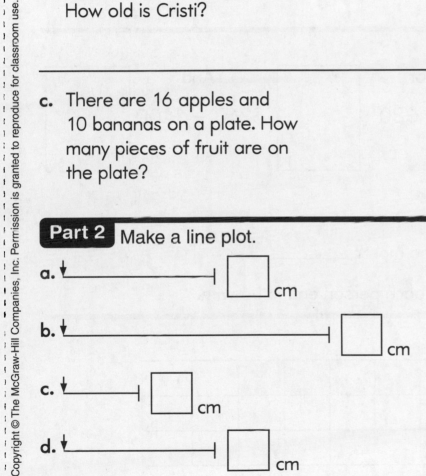

a. ☐ cm

b. ☐ cm

c. ☐ cm

d. ☐ cm

e. ☐ cm

f. ☐ cm

g. ☐ cm 0 1 2 3 4 5 6 7 8

Mastery Test 13 Name _____

Part 3 Make number families to show the different ways we can put the balls into 2 boxes.

 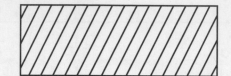

Part 4 Work each item.

Riki

$260 – $90

Junior

$260 – $80

Ana

$260 – $100

a. Who spent the most? _____

b. Who ended up with the most? _____

c. Figure out how much each person ended up with.

Part 5 Answer each question.

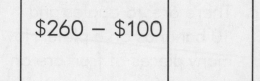

	0	1	2	3	4	5	6	7	8	9	10	11
shelf A												
shelf B												
shelf C												

books

a. How many books are on shelf B? _____

b. Which shelf has the most books? _____

c. How many more books are on shelf C than on shelf B? _____

88 Mastery Test 13

Connecting Math Concepts

Mastery Test 13

Name _____

Part 6

| a half | a third | a fourth | a fifth |

 a.

How many parts are there? _____

What is each part called? _____

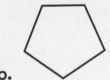 b.

How many parts are there? _____

What is each part called? _____

Part 7   Work the addition problem for each item. Write the unit name.

a. About how many feet is 4 meters? _____

b. About how many feet long is the bicycle? _____

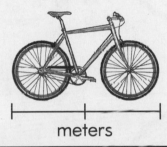

meters

Part 8   Write a letter inside each shape.

> P—Pentagon
> H—Hexagon
> Q—Quadrilateral

a.

b.

c.

d.

e.

f.

Part 9   Write each answer.

a. 2 + 56 = _____

b. 98 − 7 = _____

c. 75 + 5 = _____

d. 66 − 3 = _____

Mastery Test 13   Name _____

Part 10 Work the addition problem and the times problem for the figure.

a. $+$

b. _____

Part 11 Figure out how many hours go by.

a.

AM

PM

b.

PM

PM

_____   _____

c.

PM   AM

_____

Part 12 For each row, circle odd or even. Then write the number of objects.

a. 🍄🍄🍄🍄🍄🍄🍄🍄🍄🍄🍄   odd / even _____

b. 🗼🗼🗼🗼🗼🗼🗼🗼   odd / even _____

c. 🍦🍦🍦🍦🍦🍦🍦🍦🍦🍦🍦🍦   odd / even _____

90 Mastery Test 13

Connecting Math Concepts

Mastery Test 13 Name _____

Part 13 Write each problem and figure out the answer.

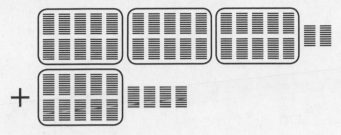

a. _____ + _____ = _____

Subtract 140

b. _____ – _____ = _____

Part 14 Show each problem on the number line. Then write the answer.

a.
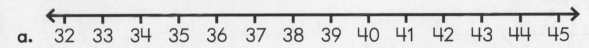
32 33 34 35 36 37 38 39 40 41 42 43 44 45

38 + 5 = _____

b.
53 54 55 56 57 58 59 60 61 62 63 64 65 66

61 – 4 = _____

Remedies

Name _____

Part A

a.  ☐ cm

b. ☐ cm

c. ☐ cm

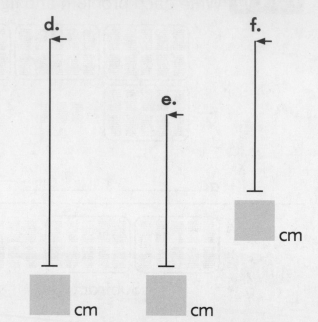

d. ☐ cm

e. ☐ cm

f. ☐ cm

0 1 2 3 4 5 6 7 8

Part B

a. ☐ cm

b. ☐ cm

c. ☐ cm

d. ☐ cm

e. ☐ cm

f. ☐ cm

g. ☐ cm

h. ☐ cm

0 1 2 3 4 5 6 7 8

Remedies CONTINUED

Name _____

Part C

a.

Sam

$66 − $65 _____

Fran

$66 − $62 _____

Kay

$66 − $63 _____

b.

Rob

$50 − $12 _____

Cal

$50 − $15 _____

Jan

$50 − $16 _____

Part D

a.

May

$99 − $55 _____

Bob

$99 − $65 _____

Ray

$99 − $75 _____

b.

Rob

$350 − $190 _____

Cal

$350 − $150 _____

Jan

$350 − $180 _____

Part E

	0	1	2	3	4	5	6	7	8	9	10	
closet A												_____
closet B												_____
closet C												_____
closet D												_____

Number of Coats

1. Which closet has the most coats? _____

2. Which closet has the fewest coats? _____

3. Which closets have more than 3 coats? _____

Remedies

Remedies CONTINUED

Name _____

Part F

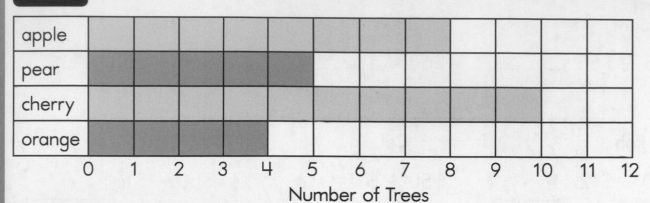

	0	1	2	3	4	5	6	7	8	9	10	11	12
apple													
pear													
cherry													
orange													

Number of Trees

a. How many more cherry trees than apple trees are there? _____

b. How many fewer pear trees than cherry trees are there? _____

c. How many fewer orange trees than pear trees are there? _____

Part G

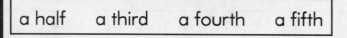

a half	a third
a fourth	a fifth

a.

How many parts are there? _____

What is each part called? _____

b.

How many parts are there? _____

What is each part called? _____

c.

How many parts are there? _____

What is each part called? _____

d.

Part H Answer each question.

a half	a third	a fourth	a fifth

a.

How many parts are there? _____

What is each part called? _____

b.

How many parts are there? _____

What is each part called? _____

c.

How many parts are there? _____

What is each part called? _____

Connecting Math Concepts

Remedies CONTINUED

Name _____

Part I

a. 58 − 2 = _____

b. 58 + 2 = _____

c. 3 + 63 = _____

d. 9 + 12 = _____

e. 75 − 5 = _____

f. 2 + 84 = _____

g. 99 − 1 = _____

h. 79 + 5 = _____

Part J Write each answer.

a. 46 + 6 = _____

b. 9 + 27 = _____

c. 28 − 3 = _____

d. 4 + 13 = _____

e. 2 + 56 = _____

f. 88 − 4 = _____

g. 96 − 6 = _____

h. 95 + 5 = _____

Part K

a.
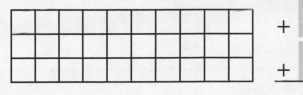

b.
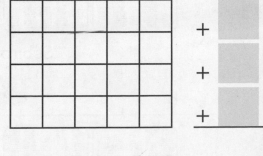

Remedies CONTINUED

Name _____

Part L

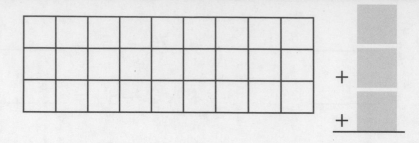

Part M

a. □ □ | □ □ □ □ □ □ □ □ □   odd / even _____

b. ▽ ▽ ▽ ▽ ▽ ▽ ▽ ▽ ▽ ▽ ▽ ▽ ▽   odd / even _____

c.    odd / even _____

Part N

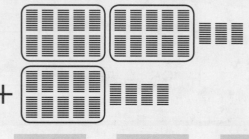

a. ▢ + ▢ = ▢

Subtract 120

b. ▢ − ▢ = ▢

Subtract 350

c. ▢ − ▢ = ▢

Connecting Math Concepts

Remedies

Remedies CONTINUED

Name _____

Part O

a.
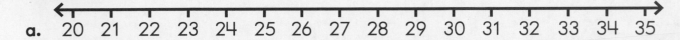
20 21 22 23 24 25 26 27 28 29 30 31 32 33 34 35

$$27 + 6 = \underline{\qquad}$$

b.
60 61 62 63 64 65 66 67 68 69 70 71 72 73 74 75

$$68 + 4 = \underline{\qquad}$$

c.
46 47 48 49 50 51 52 53 54 55 56 57 58 59 60

$$53 - 6 = \underline{\qquad}$$

Part P

a.
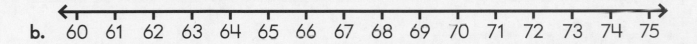
33 34 35 36 37 38 39 40 41 42 43 44 45 46 47 48 49 50

$$42 - 8 = \underline{\qquad}$$

b.
12 13 14 15 16 17 18 19 20 21 22 23 24 25 26 27 28

$$19 + 5 = \underline{\qquad}$$

c.
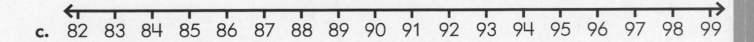
82 83 84 85 86 87 88 89 90 91 92 93 94 95 96 97 98 99

$$90 - 7 = \underline{\qquad}$$

Remedies

Cumulative Test 2

Name _____

Write each answer.

a. 6 + 5 = _____	**f.** 10 – 4 = _____	**k.** 5 + 6 = _____	**p.** 10 – 6 = _____
b. 14 – 7 = _____	**g.** 15 – 8 = _____	**l.** 6 + 8 = _____	**q.** 11 – 6 = _____
c. 13 – 5 = _____	**h.** 8 + 5 = _____	**m.** 4 + 8 = _____	**r.** 14 – 8 = _____
d. 6 + 6 = _____	**i.** 6 + 4 = _____	**n.** 13 – 7 = _____	**s.** 8 + 7 = _____
e. 7 + 6 = _____	**j.** 12 – 8 = _____	**o.** 8 + 8 = _____	**t.** 12 – 6 = _____

Part 2 Write the letter or letters inside each shape.

C: Circle	Cu: Cube	H: Hexagon	P: Pentagon
Py: Pyramid	Q: Quadrilateral	R: Rectangle	RP: Rectangular Prism
	S: Square	Sp: Sphere	T: Triangle

a.

b.

c.

d.

e.

f.

g.

h.

i.

j.

k.

Cumulative Test 2 Name _____

a. M is 12 less than P.
P is 58.
What number is M?

b. V is 30 less than X.
V is 43.
What number is X?

c. T is 27 more than R.
T is 99.
What number is R?

Part 4 Answer each question.

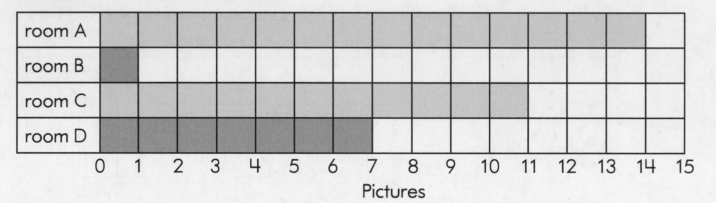

Pictures

a. Which rooms have more than 8 pictures? _____

b. How many pictures are in room A? _____

c. Which room has the fewest pictures? _____

d. How many more pictures are in room C than room D? _____

Connecting Math Concepts

Cumulative Test 2

Name _____

Part 5 Work each problem.

a. Tami is 10 years younger than Fabi. Tami is 14 years old. How many years old is Fabi?

b. Jeff started out with some dollars. He spent $42. He has $39 left. How many dollars did Jeff start out with?

c. There are 289 cars on the lot. If 125 of the cars are new, how many used cars are on the lot?

d. There are 99 girls and 57 boys at the fair. How many fewer boys than girls are at the fair?

Part 6 Work the times problem for each item.

a.  _____

b. How many cents is 2 quarters? _____

c. How many cents is 6 nickels? _____

Cumulative Test 2

Name _____

Part 7 The table shows the sizes of colored shirts in a store. Answer each question.

	small	medium	large
red	8	14	20
yellow	25	2	18
blue	11	15	10

a. How many large shirts are blue? _____

b. The fewest small shirts are what color? _____

c. 14 red shirts are what size? _____

d. How many medium shirts are yellow? _____

Part 8 Work each item.

$14.95 $11.50 $12.25 $9.45

a. You buy the socks and the gloves. How much money do you still have?

b. You buy the hat, the gloves, and the scarf. How much money do you still have?

Part 9 Measure the line to each arrow.

```
 0   1   2   3   4   5   6
 0 1 2 3 4 5 6 7 8 9 10 11 12 13 14 15
```

☐ in.   ☐ cm   ☐ cm   ☐ in.
  a.   b.   c.   end

Connecting Math Concepts

Cumulative Test 2

Name _____

Part 10 Write the sign >, <, or =.

a. 99 centimeters ☐ 1 meter

b. 3 feet ☐ 1 yard

c. 1 week ☐ 6 days

d. 1 day ☐ 24 hours

e. 101 cents ☐ 1 dollar

f. 59 minutes ☐ 1 hour

g. 12 months ☐ 1 year

h. 62 seconds ☐ 1 minute

i. 1 gallon ☐ 3 quarts

j. 13 inches ☐ 1 foot

Part 11 Work each problem.

a. $2 \times \underline{\quad} = 10$

b. $5 \times 7 = \underline{\quad}$

c. $4 \times 6 = \underline{\quad}$

d. $9 \times \underline{\quad} = 27$

Part 12 Complete each equation.

$\underline{\quad} + \underline{\quad} = 55$

$79 + \underline{\quad} = \underline{\quad}$

a.

b.

Part 13 Circle **odd** or **even** for each number.

a. 9   odd / even

b. 30   odd / even

c. 12   odd / even

d. 21   odd / even

Cumulative Test 2

Name _____

Part 14 Write the sign >, <, or =.

a. 24 ☐ 204

b. 239 ☐ 200 + 40 + 1

c. 40 + 9 ☐ 50

d. 128 ☐ 127

Part 15 Write the time for each clock.

a. _____ b. _____ c. _____ d. _____

Part 16 Write the column problem for each item.

a. 24 + ☐ = 54

b. ☐ – 15 = 29

Part 17 Write each dollars and cents number: $■■.■■

a.

   _____

b.

     _____

Cumulative Test 2

Name _____

Part 18 Answer the question. Remember the unit name.

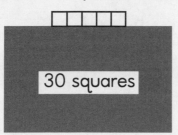

30 squares

How many rows are there?

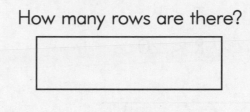

_____

Part 19 Write the estimation problem and the answer.

a. 2 6
   + 1 2
   ‾‾‾‾‾

b. 4 3
   + 5 1
   ‾‾‾‾‾

c. 9 2
   − 7 8
   ‾‾‾‾‾

Part 20 Find the perimeter of each figure.

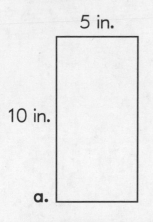

5 in.

10 in.

a.

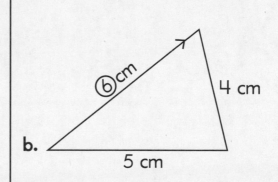

⑥ cm

4 cm

5 cm

b.

Part 21 Find the area of the rectangle.

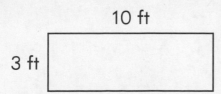

10 ft

3 ft

Cumulative Test 2   Name _____

a. $T < R$
    $R < 20$

_____

b. $20 > M$
    $V > 20$

_____

Part 23 Write each answer.

a. $60 - 20 =$ _____

d. $61 + 8 =$ _____

g. $73 + 9 =$ _____

b. $90 + 40 =$ _____

e. $38 + 2 =$ _____

h. $280 - 100 =$ _____

c. $89 - 7 =$ _____

f. $51 + 10 =$ _____

i. $420 + 100 =$ _____

Photo Credits